KNOWLEDGE CAPSULE

KNOWLEDGE CAPSULE

Vol. 01

HUZEFA

PARTRIDGE

Copyright © 2016 by Huzefa.

ISBN: Hardcover   978-1-4828-4864-9
           Softcover   978-1-4828-4865-6
           eBook   978-1-4828-4863-2

All rights reserved. No part of this book may be used or reproduced by any means, graphic, electronic, or mechanical, including photocopying, recording, taping or by any information storage retrieval system without the written permission of the author except in the case of brief quotations embodied in critical articles and reviews.

Because of the dynamic nature of the Internet, any web addresses or links contained in this book may have changed since publication and may no longer be valid. The views expressed in this work are solely those of the author and do not necessarily reflect the views of the publisher, and the publisher hereby disclaims any responsibility for them.

Print information available on the last page.

To order additional copies of this book, contact
Partridge India
000 800 10062 62
orders.india@partridgepublishing.com

www.partridgepublishing.com/india

Contents

ACKNOWLEDGMENT vii

PREFACE ix

FASHION, FOOD & ENTERTAINMENT 1

GEOGRAPHY 15

ENGINEERING & TECHNOLOGY 27

HISTORY 39

SCIENCE 51

SPORTS 63

WORLD RECORDS 77

MIXED BAG 91

ACKNOWLEDGMENT

First of all, I would like to express Gratitude & thanks to our Spiritual leader without whose blessings this book would not have been possible. I would also like to thank my Parents, my wife Mariya, My Cutie-pie son & my two sweet sisters – Maryam & Fatema for their constant Support & Encouragement. Also to people who read, wrote, offered comments, allowed me to quote their remarks and assisted in the editing, proofreading and design. **SPECIAL THANKS TO PARTRIDGE PUBLISHERS** for their Constant Support & Guidance without which this book would not have been possible.

PREFACE

This book being my first shot as an Author has been obviously special. Frankly speaking, I have been an ardent "General Knowledge" Seeker since my childhood when I used to not leave any chance of gaining Knowledge go waste. I have really enjoyed a lot in taking efforts in searching facts for this book. My biggest Motivation to pen down this book was to give the lakhs of students, Teachers & Knowledge seekers like me - a Compact & Reliable source which can get into their thirsty minds easily.

So what makes this book different than the numerous Fact Books available in the market? The book has been carefully divided into various sections like Fashion & Entertainment, Geography, History, Engineering & Technology, Science, Sports, World Records & a special section named Mixed Bag containing Multiple Objective type Questions. Most of the facts are unique & innovative in their own way. The most important thing about this book is that it contains numerous Illustrations which makes it all the more interesting to read & easy to digest.

Finally I would like to conclude that there are no dearth of resources, online encyclopedias, websites etc available, but the need of the hour is a Compact Compendium which will help to enhance the knowledge & in-turn increase the IQ of the the aspiring students. In a nut-shell, there's something in this book for everyone. I just hope that people will like my humble efforts.

Suggestions are always welcome for Improvement in the next editions.

Happy Reading!!

Dedicated to my Lovely Family

FASHION, FOOD & ENTERTAINMENT

I. Fashion & Entertainment

1. ***Levis Jeans*** *was invented in **1873** by **Levis Strauss & Jacob Davis**. So it is probably one of the oldest jeans & one of my favourites!!*

2. ***Paris*** *is called as the **Perfume Capital** of the World. So, the next time anybody are off to visit Eiffel Tower, also do not forget to get hold of a Givenchy or a Calvin Klein!!*

3. *Actress **Meryl streep** holds the record for the most **Oscar Nominated Actress with 13 nominations**. What an actress!!*

4. *The **Nike Swoosh** was invented by **Caroline Davidson** in 1971 and the **first shoe** was introduced in **1972**.*

5. ***Versace*** *presented the 1ˢᵗ Designer Collection in **1978**.*

6. ***Milky-way Chocolate*** *was launched in **1922**. Yummy!!*

7. ***Bernd Eilts***, *a German artist turns dried **cow manure** into wall clocks and small sculptures. He has now expanding his business to include Cow dung Wrist watches! Now that's certainly something Creative & Unique!!*

8. ***Louis Vuitton*** *was founded in 1854. The name is enough!!*

9. *Food Brand - **Pizza-hut** was opened in 1958.*

10. *Famous Pop Band group got its name **ABBA** by taking the first letters from each of the names of their members (Agnetha, Bjorn, Benny and Anni-frid)*

11. *The first modern **Brassiere** was invented by New York Socialist **Mary Phelps** in 1913 using two Silk Handkerchiefs & a ribbon.*

12. ***Reita Faria** was the first Indian Woman to win Miss World crown in 1966. We all are so proud of this splendid achievement Reita!!*

13. *At the age of just 4 years, **Mozart** was able to learn a piece of Music in half an hour! No wonder that he is called the **"King of Music"** with over 600 compilations in all genres of music!!*

14. ***Walt Disney** has won the most no. of Oscars till now by an individual. He was nominated for 64 and won 26!*

15. ***"The card Players" Painting** made by Paul Cezanne in 1892 is the highest paid painting. It was sold for **267 Million $** in April 2011 and bought by The Royal family of Qatar!*

16. *In 1962, the first **Wal-mart** store opened in Rogers, Arkansas, USA. At present it has over 11000 stores in 28 countries with over 2.2 million employees! Well, no competition with this giant in this field!!*

17. ***Hallmark** makes cards for 105 different relationships with more than 800 artists, designers, stylists, writers etc and producing more than 19000 new & re-designed cards per year.*

18. *Actor **John Travolta** is also a **Certified Private Jet Pilot** and has his own airfield and fleet of 5 planes including a Boeing 707! Now that's a Multi-talented personality!*

19. ***Red-bull drink** was invented in 1987 by an Austrian company Red bull GmbH.*

20. *There was no **Punctuation** in English until 15th Century.*

Paris is called as the Perfume Capital of the World.

21. *Before they became famous in their respective acting careers,* **Johny Depp** *worked as a Salesman,* **Keanu Reeves** *worked at a Pasta shop in Toronto &* **Jack Nicholson** *worked as an Office boy in MGM's Cartoon department! As they say, "Nobody knows what any person has gone through in their lives, to achieve Success" & these successful celebrities certainly prove it all the more!*

22. **Charlie Chaplin** *became the first Hollywood Actor to appear on the cover of the Time's Magazine in 1925. An Eternal Legend…*

23. *The biggest* **Bell** *is the* **"Tsar Kolokol"** *cast in Moscow Kremlin, Russia in 1733 weighing* **216 tons**. *It was broken during casting and has never been rung. It was commissioned by Empress Anna Ivanovna, niece of Peter the great.*

24. **Bud Spencer** *is an Italian film* **Actor** *who is known to be Multi-talented. He was known to be an excellent* **Swimmer** *having being the 1st Italian to swim 100m free-style in less than a minute. He also also a Jet airplane & Helicopter* **Pilot**, **Writer and Music Composer!!** *Now that's a hell of a Mutli-talent!!*

25. **"Batman"** *created by Bob Kane & Bill Finger in 1939 has been declared the* **Most Popular Super-hero** *of all time through various Surveys & Polls. Certainly, the Dark knight ought to be coz of the tragedy that captures him during his childhood & how he trains himself brilliantly & rises to be a Savior of Gotham city without having any Special abilities like the other Super-heroes!!*

26. *The Piccolo is the smallest of all* **Wood-wind Musical Instruments.**

27. **Calvin Klein** *was found in* **1968** *by an American Fashion designer with the same name. The Company is headquartered at New York. It was initially opened as just a Coat shop.*

28. **Koumis**, *an alcoholic Beverage made from fermented mare's milk is the national drink of* **Mongolia.** *Would really love to give it a try!!*

29. *In 1980, Namco released* **PAC-MAN**, *the most popular Video game of all time. The original name was "PUCK-MAN" but was changed later.*

30. Famous Novelist **Stephen King's** 1st Novel was **"Carrie"** published in 1973.

31. The **National Dish of Greece** is **Moussaka**- a divine meal prepared with eggplant, ground beef, onions, tomatoes, spices and béchamel sauce.

32. **The first Starbucks Coffee shop** opened at Seattle, USA in 1971 by three friends from University of San Francisco – History Teacher- Zev Siegl, English teacher- Jerry Baldwin & Writer – Gordon Bowker. It was initially named as Pequod but later rejected. Thanks to the trio for giving us such a wonderful place to spend some quality time with our loved ones!!

33. **Mandarin Chinese** is the **most widely spoken language** in the world with **14.1 %** of the world's population speaking it. And I thought English is the most widely spoken one!!

34. **"Dark Side of the Moon"** (a Pink Floyd album) stayed on the top 200 Billboard charts for **741 weeks of 14 years**!

35. During his Lifetime, **Vincent Van Gogh** sold just one of his Paintings – **The Red Vineyard** out of his 900 paintings! Now that's the best example for some serious dedication towards one's work!!

36. **Kevin Costner** won an Oscar for **"Dancing with the Wolves"** in 1990.

37. **Roger Moore** has played maximum times (7 times) as James Bond in **James Bond Movies**.

38. **"Monopoly"** was first invented by American Salesman **Charles Darrow** in 1935. That really reminds me of my childhood memories with my siblings!! My Favorite Board-game!!

39. **Caesar Salad** was first prepared in 1924 by **Caesar Cardini**, an Italian American Chef & Restaurateur.

40. Popular American Rock Band **"Linkin Park"** was formed in 1996. It rose to fame after its **1st Album- Hybrid Theory** was a smashing hit. An Awesome & rocking song it was!!

41. **Gaumont Film Company** is a French film production and distribution company founded by the engineer-turned-inventor **Léon Gaumont** (1864–1946). It is the first and **oldest continuously operating film company** in the world, founded before other studios such as Nordisk Film, founded in 1906, and Universal Studios and Paramount Pictures, which were both founded in 1912.

42. **Entertainers** who worked in the food business before they became famous include **Stephen Baldwin**, who was a **Pizza Parlour** Employee, **Bill Murray** was a Pizza maker, **Jean Claude Van demme** delivered Pizzas. Many years ago **Julia Roberts & Christie Brinkley** both sold **Ice-cream**. Before she made it as a pop singer, **Madonna** sold **doughnuts** at Dunkin Donuts. **Jennifer Aniston** was a **waitress** at a Burger joint!!

43. In **Japan**, by the time man reaches the age of 60, he is **commemorated** with a special Ceremony. This Ceremony features the man wearing a **red Kimono**, which denotes that he no longer has the responsibilities of being a mature adult.

44. The **first product** to ever be scanned with a **bar code** was **Wrigley's gum** on **June 26, 1974**.

45. **Blue-Berries** have more **anti-oxidants** than any other Fruits & Vegetables.

46. **Heinz Tomato Ketchup** was introduced in **1876** by **F&J Heinz Co.** (Frederick & John Heinz). Certainly the most delicious & juicy ketchups amongst all the brands!!

47. The **Frankeinstein Novel** was first published by **Mary Shelley** in 1818 in London.

48. The Current **Sultan of Brunei** is Sultan Haji Hassan Al Bolkiah. He is also the 1st incumbent Prime-minister of Brunei. One of the last remaining absolute monarchs in the world, Sultan has earned enough from reserves of oil and natural gas, that he doesn't have to think twice about getting a haircut for a mindboggling **$ 21,000**! His personal collection of over 7,000 high performance cars which by some estimates is worth more than $ 5 billion includes 600 Rolls Royce cars, over 300 Ferraris, 134 Koenigeggs, 11 McLaren F1s, 6 Dauer Porsche 962 LMs and a number of luxurious Jaguars. He is famous for having some of the most luxuriously customized private jets like Boeing 747-400 and Airbus 340-200!! His **net worth is 19.5 billion $**

49. In the game **Monopoly**, the **most money** one can lose in one travel around the board (normal game rules, going to jail once) is **$ 26,040**. The most money you can lose in one turn is **$ 5070**.

50. In her entire Lifetime, **Spain's Queen Isabella** (1451-1504) bathed just twice!! Now that's because bathing was prohibited during the Medieval ages due to various superstitious beliefs by the Church! Poor lady!!

51. **The largest Single-screen Movie Theatre** in the world was Gaumont Palace in Paris, France with a seating capacity of 6,420. The Palace was closed in 1970 & demolished in 1972 leaving New York's 5933 seat Radio City Music theatre as the largest one. **The largest megaplex** in the world by **screens** is the **AMC Mesquite in Mesquite**, **Texas** with **30 screens** and a total seating capacity of **6,008**. **The largest megaplex in the world by seats** is Metropolis in Antwerp, Belgium, with 24 screens and a total **seating capacity 0f 8,092.**

52. Leading Electronic Company **"Sony"** was founded in 1946. It was originally called **"Totsuken"**. It was felt that the name "Sony" would be easier to pronounce. The name was invented by a cross between the name "sonus" and "sonny".

53. In the movie **"Speed" (1994)**, **twelve buses** were used, including two which exploded, one for the free-way jump; one for high speed scenes; and one solely for under bus shots. Now that's called Intelligent film making!!

54. **The Liberace Museum** at Nevada, **Las Vegas** opened in 1979 by **American Pianist & Vocalist Valentino Liberace** had a mirror plated **Rolls Royce**; jewel-encrusted capes, and the largest rhinestone in the world, weighing 59 pounds and almost a foot in diameter.

55. **Cherophobia** is the fear of Fun!! Really have to pity the person who do suffer from such kind of conditions after all Life is to enjoy & have fun!!

56. The **world's first motel** opened in **San Luis Obispo, California** in **1925**. It was initially called "Milestone Motel" & later changed to **"Motel Inn"**.

57. **King Kong (1933)** was **Adolf Hitler's Favorite movie.** No wonder he got inspired to carry out destruction & annexation!!

The National Dish of Greece is Moussaka- a divine meal prepared with eggplant, ground beef, onions, tomatoes, spices and béchamel sauce.

58. **Chicago, Illinois** is the **candy capital** of the world. Chicago has more **chocolate manufacturers** within a small radius than any other place in the world. This dates back to the 1800's when Chicago was a **national hub for transportation - & manufacturing** in addition to being very close to sources for key candy. It was also convenient to ship Candy products to either coast from Chicago. Wow!! One must visit the Candy capital!!

59. **Meg Ryan** turned down plum lead roles for the films "Steel Magnolia", "Pretty Woman" & **"The Silence of the Lambs"**, which earned **Jodie Foster** a **Best Actress Oscar**. That's sooooo unlucky Meg!!

60. **Earl Dean** an employee of **Root Glass Company** designed the bottle for **Coca Cola Co.** in **1915**.

61. The **oldest actor** to win a **Best Actor Oscar** is **Henry Fonda** for the movie "On Golden Pond" in **1981**. Such a fantastic actor!!

62. **George W. Bush and Saddam Hussein** had their **shoes** hand-made by the same **Italian** cobbler! No wonder they were connected so closely!!

63. On September 9, 1950 **dubbed laughter** was used for the **first time on television**. It was used for the sitcom **"The Hank McCune Show"**.

64. **John Kelloggs** an American Medical Doctor in Battle Creek, Michigan invented **corn flakes**, for a patient with bad teeth in **1906**. Till today, it's the staple diet for breakfast for so many Americans!!

65. **The first Television advertisement** was broadcast in the United States at 2:29 p.m on **July 1, 1941**. It was aired at the **10 second** Spot before a Baseball game between Brooklyn Dodgers & Philadelphia Phillies showing a Static **Bulova Watch** ticking over the United States Map.

66. **Sherlock Holmes** is the **most portrayed character** on **film**, having been played by **72 actors in 204 films**! The historical character most represented in films is of **Napoleon Bonaparte**, with **194 film portrayals**. Thus proving that detective characters are still the most popular & in demand ones!!

67. **In Australia**, the Number 1 **topping for pizza is eggs**. In **Chile**, the **favorite topping is mussels & clams**. In the **USA, it's Pepperoni**.

GEOGRAPHY

II. Geography

1. ***Rufiya*** is the Currency of ***Maldives.***

2. *The City of **Las-vegas** has the most no. of Hotel Rooms in the world. That's gotta be a big achievement beating the likes of London, New york, Paris etc!!*

3. *The driest place in the world is **CALAMA in the Atacama Desert** in Chile.*

4. ***Quito, Ecuador*** *is the city situated closest to **The Equator**. Not surprising at all for naming the city lying near the line dividing the earth into two equal halves after it!*

5. ***Sydney Harbour****, Australia is the World's biggest Natural Harbour.*

6. ***Hawaii's Mount Waialeale*** *is the **wettest place** in the world where it rains about 90% of the time about 480 inches per annum. Oh that must be really tough for the residents around that area!*

7. *The **most densely Populated City** in the world is **Hong-kong**, China. Also, at the same time it is one of the most expensive cities in the world!!*

8. ***Puma*** *is the national animal of **Argentina**.*

9. ***Jasmine*** *is the national flower of Pakistan.*

10. *The **Shoreline of the Dead-sea** on the border of Israel & Jordan is the lowest point on earth at a depth of 400 meters below sea level.*

11. **Son-doong Cave** in Vietnam is the biggest cave in the world with total length measuring **9kms**! That's enough to engulf a small city!!

12. Despite being over 27 times smaller, **Norway**'s Coastline is longer than USA's! Something certainly worth boasting of!!

13. **The Pentagon** spread out at a floor area of **600,000 m2** and having a working Population of 30,000 employees operates like a small city. It has its own Shopping mall, Bank, Power plant, water & Sewage facilities, fire station, Police Force, fast-food restaurants and a Mayor! Also, the most secured office area in the world!!

14. **"Ice Hotel"** is the world's largest hotel made of Ice & Snow and is located in Sweden. It certainly is a modern age marvel and an Inspiring art-work!

15. **Nauru**, a tiny island near Australia is the only **Country** in the world without a Capital and is also the World's smallest Republic. **Yarken** is the only district there where all the important Government buildings are situated.

16. **Finland** has 187,888 lakes and 179,584 islands!! That's the reason why Finland is called the "Land of a thousand lakes"!!

17. The deepest Lake in the world is **Lake Baikal** in Russia which has a depth of 1620 mtrs.

18. **Tibet** is the **highest country** in the world. Its average height above sea level is 4500 mtrs.

19. **Chris Hani Baragwanath Hospital** at Johannesburg, South africa built on **174 acres of land with 6760 staff** and employees & 3200 beds is the **largest hospital** in the world.

20. The World's richest country is **Qatar** as of **2013** with its **GDP at Purchasing Power Parity (PPP) per capita** amounting to **$ 98,914** as per data from **International Monetary Funds (IMF)**. I would love to work there!!

21. **Macau** is the country having the highest Population Density in the World with **19,610 persons/sq. km.**

22. In some of the dry valley regions of **Antarctica** like the Mc Murdo Valleys, it has been predicted by scientists that it has never rained in 2 million years! That explains the lack of vegetation on it.

23. **The Salar de uyuni** is the **largest salt flatland** in the world at **4,086 square miles (10,582 kilometers)**. Every year, this amazing wonderland in southwest Bolivia covers with a thin layer of water. When that happen it turns into the **largest mirror** on the planet! That's nature at its best!!

24. **Kazakhistan** is the World's **largest Landlocked country** with an area of 2,722,4900 km2. That's really harsh with no access to any water resources!

25. **Limnologist** is a person who studies Lakes & Rivers. How can a person just study Lakes & Rivers!??

26. **Madagascar** is the oldest Island in the World having settled by Austronesian people between 350 BC – 550 BC. I think it's a really enchanting island with the most diverse groups of Flora & Fauna in the world.

27. The highest Capital city in the world is **La Paz of Bolivia** with an elevation of about **5099 mtrs.** Actually, Bolivia has got two capitals - first is La Paz which is the administrative capital & Sucre is the other capital. Thus, if La Paz is considered as the capital, then it is ranked no. 1 else Quito which is the capital of Ecuador grabs the top spot.

28. **Antarctica** was first discovered by **Captain James Cook** and his crew aboard "The HMAS Resolution" & "Adventure" on January 17, 1773. Actually he came close to the continent but was prevented from moving ahead due to large sheets of ice. The International **Dialling code** for Antarctica is **672**.

The Salar de uyuni is the largest salt flatland in the world at 4,086 square miles (10,582 kilometers).

29. The **Netherlands** has the highest no. of **Museums (1000 nos.)** in the World with Amsterdam alone having 42 of them!

30. **Malta** is the **Laziest country** in the world with **71.9 %** of the Population as inactive! From what I can make out, the reason for it is due to the small size of the country and also because of the high standard of the living of the people, every single soul must be having a vehicle to traverse around such a small area even!

31. **OPEC (Organization of the Petroleum Exporting Countries)** was founded at the Baghdad Conference in **1960** by Iraq, Venezuela, Kuwait, Saudi Arabia & Iran. At present it has Headquarters at Vienna, Austria since 1965.

32. The **first Sky-scraper** in the the world was built in **Chicago, USA** in 1885. It was called the **"Home Insurance Building"** running upto a height of 138 feet high and consisted of 10 stories. Unfortunately it was demolished in 1931. Nowadays, Sky is the limit for the construction of Sky-scrapers!!

33. **The Krubera Cave** located in Abkhazia district of Georgia is the deepest cave in the world with its depth measured upto **7208 fts.**

34. Next to Warsaw, **Chicago** has the highest **Polish** population in the world. Name a single nationality which doesn't have a good amount of population in the U.S!!?

35. Of all countries, **Brazil** has the largest **plant species** in the world - **56,000!!** Not surprising at all, with the Amazing Amazon-rainforest situated there!!

36. **Pakistan** has the world's largest **Canal based Irrigation system**. That's a decent achievement for a country mauled by tragic incidents in recent times.

37. **Punta Arenas** in Chile is the world's **Southernmost** city.

38. The city with the **least Air-Pollution** is **White-horse** in Yukon, Canada having an **annual average of 3 micrograms of PM10 Particles per cubic metre**.

39. **Shortest Intercontinental Commercial Flight** in the world is from Gibraltar (Europe) to Tangier (Africa). **Distance was 34 miles, flight time - 20 minutes**.

40. **Tasmania** is said to have the world's **cleanest** air. I have been dying for some fresh clean air since long!!

41. **The Capitol Building in Washington D.C** which is the meeting place of the United States of America Congress has **365 steps** representing the 365 days of the year.

42. **The German Bundestag** or Parliament established in 1949, has 672 members and is the world's **largest elected legislative body**.

43. Belgian missionary **Louis Hennepin** observed and described the **Niagara falls** first, in 1677, while he was exploring the area with Rene Robert Cavalier.

44. **The Sahara Desert** is expanding **half a mile** south every year. Well, that might be due to the changing precipitation patterns in the north of Africa.

45. **New South China Mall** in Dongguan, China is the **largest mall in the world** based on **gross area (659,612 m²)** while comes to second position when compared by total area(Dubai Mall is first). As per my knowledge, It is more than twice as big as the largest mall in the US, the Mall of America.

46. The **Spring Temple Buddha** is a statue depicting Vairocana Buddha **located in the Zhaocun township of Lushan County, Henan, China. At 128 m (420 ft)**, which includes a 20 m (66 ft) lotus throne, it is the **tallest statue in the world**. It derives its name from nearby hot spring, which is renowned for its curative properties. Simply Divine!!

47. The **deepest point on the Earth's surface** is in the Pacific Ocean located in the **Marianas Trench**. This point is called the "Challenger Deep" and is **35,818 feet deep**. If Mount Everest was kept under water, still it would not have reached the lowest point of Marina Trench!!

48. The **largest Volcano** known is on **Mars – Olympus Mons, 370 miles wide & 79000 feet high** and is almost three times higher than Mount Everest!

49. **The State of Michigan** in the US has the **longest freshwater shoreline** in the world of length **3224 miles**.

50. **The Rainbow Bridge** located at the base of the Navajo Mountain in Utah, USA is the **World's largest & Highest Natural Bridge** with a vertical span of 84 mtrs & Horizontal Span of 13 mtrs. God's own Marvel!!

51. The **largest Enclosed Water-body** on earth is the **Caspian Sea**.

52. The **Royal House of Saudi Arabia** was founded by **Mohammed Saud** & his descendants and consists presently of about **15000 members** of which Wealth & Power is possessed by **2000** of them, off course with the high rate of fertility prevalent there!!

53. The state of **Oregon** has one city named **Sisters** and another called **Brothers**. Sisters got its name from a nearby trio of peaks in the Cascade Mountains known as the Three Sisters. Brothers was named as a counterpart to Sisters.

54. The volume of **water in the Amazon river** is greater than the next **eight largest rivers** in the world combined! It is responsible for about **one-fifth** of the World's fresh water that flows into oceans. Truly the Queen of Rivers!!

55. The **World's worst Earthquake** occurred in the Eastern Mediterranean in July 1201, killing over **one million** people that were predominantly in **Egypt & Syria**.

The Rainbow Bridge located at the base of the Navajo Mountain in Utah, USA is the World's largest & Highest Natural Bridge with a vertical span of 84 mtrs & Horizontal Span of 13 mtrs.

56. There are more Irish in **New York City** than in Dublin; more Italians in New York City than in **Rome, Italy**; and more jews in New York City than in **Tel Aviv, Israel**!

57. When explorers first arrived in **Venezuela** in 16th century, they were reminded of Venice. They named the country **"Little Venice"**, which translated into Spanish is Venezuela.

58. The **World's largest art gallery** is the **Winter Palace & Hermitage** in **St. Petersburg, Russia**. Visitors would have to walk 15 miles to see the 322 galleries which houses nearly **3 million works** of art!! Heaven for Art-lovers & creative people!!

59. **The deepest hole ever drilled by man** is the **Kola Super deep Borehole**, in **Russia**. It reached a depth of **12,261 meters** (about 40,226 feet or 7.62 miles). It was drilled for scientific research and gave up some unexpected discoveries, one of which was a huge **deposit of hydrogen** – so massive that the mud coming from the hole was "boiling" with it.

60. There is a **law** still on the books in **Milan, Italy** which requires citizens to **smile** all times in public or risk a **hefty fine**! Exemptions include visiting patients in hospitals & attending funeral.

61. The **Windiest place** in the world is **Mount Washington, New Hampshire, USA**. The **highest wind speed** was on **April 12, 1934** when it reached **231 mph**.

62. There are more **pigs than humans** in **Denmark**! That explains for the huge amount of pork & bacon being exported from there!

63. **Niger (Africa)** has the **world's highest fertility rate – 7.1 children** per mother.

64. Thailand's Capital **Bangkok** means **"City of angels"**, while Malaysia's Capital **Kuala Lumpur** means **"Muddy Water"**.

65. ***Europe** is the only **continent** in the world **without a desert**!* Well it is due to a number of factors contributing to the emergence of deserts like lack of areas with scanty rain-fall, lack of rain-shadow effect or lack of cold ocean currents (in case of deserts forming near shore-lines)

66. ***The Port of Shanghai** is the **biggest port in the world** based on cargo throughout. The Chinese port handled **744 million tonnes of cargo in 2012**, including **32.5 million twenty-foot equivalent units (TEUs)** of containers. The port is located at the mouth of the Yangtze River covering an area of **3,619km²**. Shanghai International Port Group (SIPG) owns the port facility.*

ENGINEERING & TECHNOLOGY

III. Engineering & Technology

1. ***Nokia*** *got its name from a **Finnish Village**. It means* ***"The Black One".*** *And off course the "Un-broken one"!! J*

2. ***NASDAQ*** *stands for "NATIONAL ASSOCIATION OF SECURITIES DEALERS AUTOMATED QUOTATIONS"*

3. ***Toronto*** *was the first city in the world with a **Computerized Traffic signal system.***

4. *The first domain name ever registered on the Internet was **Symbolics.com**!! That's a massive achievement!!*

5. *If all the cars from the US were taken and lined up from bumper to bumper, there would be enough cars to go to the moon from earth and back!!*

6. ***Alexander Gustave Eiffel*** *designed the **Eiffel Tower** as well as the inner structure of the Statue of Liberty in New York. Such a skilled & efficient Architect he was, designing two master-pieces!!*

7. *A temperature of **70 million degrees Celsius** was generated at **Princeton University, USA** in **1978** during an experiment of Fusionism and is the highest man made temperature ever. This is another example of what not Man & Technology can achieve!!*

8. ***Nitrogen Gas*** *is used to fill **Tyres** of F1 Racing Cars. The reason for this is affiliated to the fact that nitrogen is dry & hence helps avoid corrosion/abrasion of tyres & also due to slower loss of pressure.*

9. *The World's most expensive Domain name is **Sex.com** sold to **Clover Holdings** co. for a whopping **13 million $** in 2010!!*

10. ***John Harrington**, God-son of Queen Elizabeth I invented the first **Flush Toilet** in **1596**.*

11. ***W.S Industries India limited**- a company specialized in Manufacture of Porcelain Insulators for Electrical Sub-stations, was the **first Indian Company** to achieve **ISO Certification** in 1991. Nowadays, it is the benchmark for a judging a firm's ability to deliver Quality & satisfactory products or services.*

12. ***YAHOO's** full form is "Yet another Heirarchical Officious Oracle".*

13. ***Tetris** was invented/designed by Alexander Pajitnov, USSR in the year 1984. Still love the game!!*

14. *The first **Computer Mouse** was invented in 1964 by **Doug Engelbart** and was made of Wood.*

15. ***Chrysler** built B-29 Bomber planes that bombed Japan in 1942; **Mitsubishi** built the A5M Zeroes that tried to shoot them down. Both companies now build Cars in a joint Plant called "Diamond Star". That's a sweet coincidence for both these Automobile giants!*

16. ***Formula Rossa at Ferrari World**, Abu Dhabi is the World's Fastest Roller Coaster accelerating upto a speed of 240 km/hr! That definitely gets some Adrenaline rushing in your veins!!*

17. *The **Biggest Nuclear Test** ever was done in Russia on October 30, 1961 known as **"Tsar Bomba"** over Mityushikha Bay, nuclear testing range. It yielded **57 Megatons** – equivalent of 10 times of all explosives detonated during World war II thus remaining the largest artificial explosion ever and the mushroom cloud reached till the height of 64kms!!*

18. *The **longest freestanding (supported only at the ends) escalator** in the world is inside CNN Center's atrium in Atlanta, Georgia. It rises 8 stories and is 205 feet (**62 m**) long.*

19. ***Vijay Bhatkar*** *gave India its first supercomputer, the* **PARAM 8000**, *in 1991. He did it in a record time of three years. He went on to build the* **PARAM 10000** *in 1998, one of the world's largest supercomputers, propelling India in to the group of elite five nations that possess this technology. No wonder this country produces some of the most intelligent & skilled professionals.*

20. **HP, Google, Apple & Microsoft** *have 1 thing in common apart from the fact they are IT cos – they all were* **started at Garages**! *That's the Mother of all coincidences!!*

21. **Leonardo da Vinci** *was not only a great artist but also a great inventor. All his inventions were sketched and were spectacularly ahead of time but could not be invented due to lack of scientific tools available. His notable ideas include Helicopter, Giant Crossbow, Parachute, 33 barrel machine gun etc.*

22. *The* **Viking I by NASA** *was the first Space-craft to land on the planet Mars in 1975 and stayed there till 6 years.*

The longest freestanding (supported only at the ends) escalator in the world is inside CNN Center's atrium in Atlanta, Georgia.

Knowledge Capsule

23. The World's first Internet Search engine was **'Archie'** derived from Archives and was invented by Computer students – **Alan Emtage, Bill Heelan & J.Peter Deutsch** in 1990 at Montreal.

24. **World's 1st Photograph** was taken in France in 1826 by French Scientist **Joseph Niepce**. It was a view from his Window. It took an exposure time of 8hrs. I wonder how he took it during such olden times!

25. **Edwin Howard Armstrong**, an American Electrical Engineer invented the **FM Radio** in 1933 at Columbia's Philosophy Hall.

26. There is no Nobel Prize for **Mathematics.** In my knowledge, the reason for not including Mathematics was that Alfred noble wanted to reward any people involved in "practical" inventions and he thought that Mathematics was a theoretical subject.

27. **Chevrolet co.** was founded on November 3, 1911 by **Louis Chevrolet & William C. Durant**. In 1918 it was acquired by General Motors Co.

28. The **Bell-X1** piloted by US Air force Retired Major General **Chuck Yeager** was the first Air-craft to break the sound Barrier in 1947. 29) In "Gulliver's travels", **Jonathen Swift** described the **two moon of Mars**, Phobos & Deimos, giving their exact sizes & speeds of rotation. He did this more than 100 years before either was discovered!!

29. **The Fujita Scale**, introduced by a Japanese Researcher- Tetsoya Fujita at University of Chicago in 1971 is used to measure the Intensity of **Tornadoes**.

30. **Thomas Elva Edison** known for the invention of the Light Bulb is also credited with many more amazing inventions like – the **Phonograph** in 1877, **Kinetoscope** in 1893 which eventually lead to the making of the **World's 1st Film Studio** named as "Edison Studios" in 1894! Around 1200 films were made until the studio's closure in 1918! That is the reason precisely, why he has been counted amongst one of the greatest pioneers in the field of Engineering & technology!!

31. The **first text message** was sent on 3rd December 1992, when **Neil Papworth** a 22 year old Test Engineer in UK of **"Sema Group"** send an SMS via a PC to the phone of Richard Jarvis through the Vodafone network. The message was "Merry Christmas!"

32. On December 2nd, 1942 a **nuclear chain reaction** was achieved for the first time under the stands of the **University of Chicago football Stadium**. The first reactor measured 30 feet wide, 32 feet long and 21.5 feet high. It weighed 1400 tons and the reaction produced just enough energy so as to light a small flashlight!

33. **George Eastman**, Inventor of **Photography** & the 1st Hand-held camera didn't like his picture to be taken by a camera!

34. **Satyam Online** become the *first private ISP (Internet Service Provider)* in December 1998 to offer Internet connections in India.

35. **Volkswagen's** Top selling car & the World's 2nd best-selling car- **Golf** was 1st launched in **1974.**

36. The *first car radio* on record was fitted to the passenger door of a **Ford Model T** by 18-year-old **George Frost**, President of Lane High School radio Club, Chicago, and was in use by May 1922.

37. The **Cullinan Diamond** from South-africa in **1905**, is the largest diamond found ever weighing **3106.75 carats (621 g).**

38. The **Longest & Heaviest ship** ever built was the **Sea-wise Giant or Jahre Viking** built in 1979 by Sumitomo Heavy Industries ltd. Of Japan and measured **458 mtrs** with a tonnage of **657,019 tonnes**!40) Only one **satellite** has ever been destroyed by a Meteor : the European Space Agency's **Olympus** in 1993.41) **Plutonium** created by University of Chicago Scientist Glenn Seaborg and his colleagues – was the **first man made element**. It is an unstable & highly reactive white silvery metal in appearance with a melting point of 639.5 degree centigrades & with an atomic number of 239.

39. The **Ferrari** Car factory was founded in 1929 by **Enzo Ferrari** at Maranello, Italy. He initially started racing for Alfa Romeo cars until he founded "Scuderia Ferrari".

40. The first **Type-writer** was made by **Italian Pellegrino Turri** in 1808 for his blind friend Countess Carolina Fantoni. That is something called "True love"!!

41. The **USSR** launched the **world's first artificial satellite**, **Sputnik 1** in 1957.

42. The **United States** has paved enough **roads** to circle the Earth over 150 times!

43. **The first patent for a bar code** type product (US Patent #2,612,994) was issued to inventors **Joseph Woodland and Bernard Silver** on October 7, 1952. The **first company** to produce bar code equipment for retail trade use (using UGPIC) was the **American company Monarch Marking** in 1970, and for industrial use, the British company Plessey Telecommunications was also first in 1970.

44. **Osmium** is the **densest Substance on Earth** with a relative density of **22.57 g/cm3.**

45. **Garrett Augustus Morgan** of Paris, Kentucky, has the credit of inventing **Traffic Signal**. Although many people had invented it but Garrett Morgan was one of the first to apply for and acquire a U.S. patent for an inexpensive to produce traffic signal. The patent was granted on **November 20, 1923**.

46. **Gobron-Brillié** was the name of **first car** which had achieved the **speed of 100 mph around 1904**. It was an early **French automobile** manufactured from 1898 to 1930. **Louis Rigolly** was the first person who had this achieved this speed using this car. On 17 July, in Ostende Automobile Week, Rigolly, in a 15-litre four-cylinder. Gobron-Brillie, became the first person to exceed 100 mph with an averaging of 103.56 mph per kilometer. That's a stupendous achievement for a person & car during the early 20th century!!

Volkswagen's Top selling car & the World's 2nd best-selling car- Golf was 1st launched in 1974.

Knowledge Capsule

47. **German** Black-smith & Publisher- **Johannes Gutenberg** invented the **Printing Press** in **1436**, and the **first book to ever be printed** was "The Bible". It was however in Latin rather than English.

48. **Maria Spelterini** was an **Italian tight-rope walker** who was the **only woman** to **cross the Niagara falls on a tight-rope**, which she did on **July 8, 1876**. She used two and a quarter inch wire. A true daredevil!!

49. Popular Taiwanese firm- **HTC Corporation** was founded by **Cher Wang, H. T. Cho**, and **Peter Chou** in **1997**. Initially a manufacturer of notebook computers, HTC began designing some of the world's first touch and wireless hand-held devices in 1998. And off course, at present they are equally competent in producing stylish & performance based Smart phones with rivals such as Apple & Samsung!53) There are **six million parts** in the **Boeing 747-400**. This explains the quality & the high standard set by it!!

50. On July 1962, **France** received the first **Trans-atlantic transmission** of a TV signal from a twin station in Andover, Maine, USA via the TELSTAR sattelite.

51. **Bluetooth** is named after a tenth century king of Denmark & Norway, **Harald Bluetooth**. Harald was known for uniting various warring tribes in Denmark & Norway. This Idea was proposed in 1997 by **Jim Kardach** who developed a system that would allow mobile phones to communicate with computers.56) The American Inventor **Samuel Morse**, the inventor of the **Morse code** was a **painter** as well. One of his portraits is of the first governor of Arkansas and hangs in the governor's mansion of that state. Another Multi-talented legend!

52. **DELAG** (Deutsche Luftschiffahrts-Aktiengesellschaft) was the **world's first airline**. It was founded on **November 16, 1909** with government assistance, and operated airships manufactured by **The Zeppelin Corporation with** headquarters in Frankfurt.58) In **1889**, the first **coin-operated telephone**, patented by Hartford, **Connecticut inventor William Grey**, was installed in Hartford Bank. Soon "pay phones" were installed in stores, saloons & restaurants. By **1905**, the first outdoor payphones with **booths** were installed.

53. The **Chinese** were using **Aluminium** to make things as early as 300AD. Western Civilization didn't rediscover aluminium until 1827.

54. The first **VCR (Video-Cassette Recorder)**, made in **1956**, was the size of a Piano. **Ampex** introduced the Quadruplex videotape professional broadcast standard format with its Ampex VRX-1000 in 1956. It became **the world's first commercially successful video tape recorder** using two-inch (5.1 cm) wide tape. Due to its US$50,000 price, the Ampex VRX-1000 could be afforded only by the television networks and the largest individual stations.

55. **Polytetrafluoroethylene (PTFE) or Teflon** was discovered accidently by **Dr. Roy Plunkett** at the Dupont research Laboratories during a Coolant gas experiment conducted in 1938. Teflon is the **most slippery** substance known in the world.

56. There are two **ATM Machines at Mc Murdo Station** in **Antarctica** operated by **Wells & Fargo Co**. which is capable of supporting 1258 residents. Their Maintenance is done every two years. So much good for the researchers & scientists living there!!

57. **Jean Pierre Blanchard (1753-1809)** a Frenchman was probably the first person to actually use a **parachute for an emergency**. In **1785,** he dropped a **dog** in a basket, to which a parachute was attached, from a balloon high in the air.

58. **Sony** introduced the 1st **3.5 inch floppy disk** in **1981**. I think this invention revolutionized the word "Memory" in relation to computers/machines.

59. **Lillian Gilbreth** an American Industrial Engineer was the **mother of modern management**. Together with her husband Frank, she pioneered industrial management techniques still in use today. She was one of the first "superwomen" to combine a career with her home life. A Pioneer in **ergonomics,** Gilbreth patented many devices, including an **electric food mixer**, and **trash can with step-on-lid opener**.

HISTORY

IV. History

1. *The World's most dangerous Terrorist Organization -* **Al-Qaida** *was formed in the year* **1988** *by* **Abdullah Yusuf Azzam & Osama Bin Laden** *in* **Peshawar, Pakistan.**

2. *Height of the* **Light-house of Alexandria**-*one of seven Ancient Wonders of world was 120 mtrs. It was built by the Ptolemic Kingdom between 280 & 247 BC. It crumbled down to ruin due to the occurrence of 3 earthquakes between 900-1300 A.D.*

3. **The San-Francisco Cable Cars** *are the only* **Mobile National Monuments** *in the World.*

4. **The Shortest war in history** *was between Zanzibar & England where the former surrendered after 38 minutes! Don't know why the former wasted this much precious time also in order to surrender against the latter super-power!!*

5. **The Buckingham Palace** *in UK consists of* **775 rooms** *which includes 19 state rooms, 52 Royal & Guest rooms, 188 staff Bedrooms, 92 offices & 78 Bathrooms. The Royal* **garden covers 40 acres**, *and includes a* **helicopter landing area**, *a* **lake**, *and a* **tennis court**. *It is home to* **30 different species of bird** *and more than* **350 different wild flowers**, *some extremely rare!!*

6. **John F. Kennedy** *was assassinated in 1953 as he rode in a motorcade through* **Dealey Plaza in downtown Dallas, Texas.**

7. **John Mc-donald** *was the* **1st Canadian Prime-minister** *in* **1867.**

8. *Oil Tycoon, **John D Rockefeller** was the world's **1ˢᵗ Billionaire**. He was a co-founder of the Standard Oil Company, which dominated the oil industry and was the first great U.S business trust.*

9. ***Adolf Hitler** was the one responsible in the creation of the **Volkswagen Beetle** as he wanted a car that was cheap enough for the average German working man to afford.*

10. ***James I** was the **youngest person** to become **Ruler of England** at the age of **3 months**! So much responsibility for the poor infant!!*

11. *'**Ahura Mazda**' is the name of the **Persians Chief God**.*

12. ***Tomatoes** originated from **Peru**, South America in the **16ᵗʰ Century**.*

13. ***Abdul Kassim Ismail**, Grand Vizier of Persia in the 10ᵗʰ century carried his library with him wherever he went. 400 Camels carried the 117,000 volumes!!*

14. ***Alexander the great & Julius Caesar** both were **Epileptic**. Epileptic seizures are episodes that can vary from brief and nearly undetectable to long periods of vigorous shaking & unstableness.*

15. *Each **king** in a **deck of playing cards** represents a great king from History. Spades – King David of Israel; Clubs – Alexander the great; Heart – Charlemagne/Charles I of Italy and Diamonds – Julius Caesar.*

16. ***Austria** was the first country to start using **Post-cards** in 1869.*

The Buckingham Palace in UK consists of 775 rooms which includes 19 state rooms, 52 Royal & Guest rooms, 188 staff Bedrooms, 92 offices & 78 Bathrooms.

17. **Vijaya Lakshmi Pandit** is the only **Indian Woman** to become **President of the UN General Assembly** in 1953. She was a notable diplomat & politician, being the sister of Jawaharlal Nehru & the grand-aunt of Rajiv Gandhi.

18. **Menes/King Narmer** was the **first Pharaoh of Egypt** who ruled around 3100 BC. He was killed by a Hippopotamus after ruling for 62 years.

19. **Damascus** is the **World's oldest inhabited city** dating back to 10,000 BC. The Umayyad caliphate created Damascus as its capital, setting the scene for the city's ongoing development as a living Muslim, Arab city upon which each succeeding dynasty has left and continues to leave its mark.

20. **Great Britain** was the first country to issue **Postage stamps** in 1840.21)

21. **Liberia & Ethiopia** are the only two **African Countries** that were not **colonized** ever. Italians made an attempt to have Ethiopia as their share of Africa but they were embarrassingly defeated by the Ethiopians to become the first European losers in a colonial war.

22. The Brazilian city of **Rio de Janeiro** was discovered on January 1, 1502 by Portuguese Navigators accidently. It means **"River of January"**.

23. The Historical city of **Rome** was founded in **753 BC**. According to legend, it was founded by twin boys, **Romulus & Remus**. Romulus killed Remus and named the city on him-self.

24. In 1916, **Margaret Sanger** – an American Activist, Nurse & Sex educator was jailed for 1 month for starting the **1ˢᵗ Birth-control clinic** in US.

25. The last time a **"Cavalry charge"** was used in war was in **World War II** when a **Mongolian Cavalry Division** charged against a **German Infantry Division** and the result? Not one German was killed but 2000 of the cavalry were!

26. **The Eiffel Tower** – *an architecture Marvel was designed by **Alexander Gustave Eiffel** to commemorate the 100th anniversary of the French Revolution and was completed on 31st March 1889. It consists of **18,038 Iron bars** with the total weight of Metallic & Non-metallic parts as 10,100 tons. Its total height is 324 mtrs & it also has a Post-office at its first floor. Total Construction cost was **7,799,401.31 French gold francs in 1889.***

27. *The idea of **"Earth day"** came to **Gaylord Nelson**, then a **U.S. Senator from Wisconsin**, after witnessing the ravages of the 1969 massive oil spill in Santa Barbara, California. In 1990, Earth Day became a global event. Two hundred million people around the world staged dramatic displays of environmental support, such as a 500-mile human chain in France.*

28. ***Aphrodite*** *is the Greek goddess of **Love, Beauty & Pleasure**. According to Homer's Illiad, she is the daughter of Zeus (Goddess of the sky) & Dione. Her Roman Equivalent is Venus.*

29. *In 1907, the first **taxi-cab** took to the streets of **New York City** and was started by **Harry Nathaniel Allen**. He imported 600 gas-powered taxi cabs from France.*

30. *Revealed more than 2000 years ago, the **"Zohar"** is a group of 23 Spiritual books of the **Jews** that explains the various secrets of the Universe, Bible & every aspect of life and was discovered by a revered sage named **Rav Shimon bar Yochai.***

31. *In the late 30's, a man named **Abe Pickens of Cleveland**, Ohio attempted to promote **World peace** by placing personal calls to various country leaders. He managed to contact Mussolini, Hitler, Hirohito & Franco. He spent 10,000 $ for this great purpose!*

32. *It took **20,000 men & 22 years** to build the **Taj – Mahal**.*

33. ***Leonardo da Vinci*** *could **paint** with one hand and **draw** with the other at the same time!! The reason believed for this was that his right hand was paralyzed & he had this superb gift of using his left hand as well as his right!*

34. **Louis XIV or "Louis the great"** a Monarch of France from 1661-1733 had forty personal wigmakers and almost 1000 **wigs**!

35. More than **5600 men died** while building the **Panama Canal**. Today it takes more than 8000 men to run & maintain the canal. It takes a **ship** an average of **33 hours** to travel the length of the canal!

36. **French Pilot Didier Delsalle** made the 1st ever **Helicopter landing** on the **Mount Everest** Summit on **May 14, 2005**.

37. **New Zealand** was the first country to allow **Women to vote** in **1893**.

38. **Napoleon** took **14,000 French Decrees** and simplified them into a Unified set of 7 laws. This was the **first time in modern History** that a nation's laws applied equally to all citizens. Napoleons **7 laws** were so impressive that by 1960 more than **70 governments** had adjusted their own laws according to his!

39. **France** was the first country to introduce the **license plate** with the passage of the Paris Police Ordinance on **August 14, 1893**, followed by **Germany in 1896**. The **Netherlands** was the **first country to introduce a national license plate**, called a "driving permit", in **1898**.

40. **Adolf Hitler** was the Time Magazine's **"Man of the year"** for 1938. The Time's Magazine claimed that the "Person of the year" was awarded to the public figure that had the most effect on world affairs over the year. As Time has pointed out many times since, it is not an endorsement of the person or their deeds

41. **George Washington** died after being **bled by Leeches** in order to treat "Inflammatory Quincy" in **1799**.

42. **Pope Paul IV**, who was elected on **23rd May 1555**, was so outraged when he saw the **naked bodies** on the ceiling of the **Sistine Chapel** that he ordered Michelangelo to paint on to them.

43. **Queen Supayalat**, the last queen of **Burma** ordered a **massacre** of 100 of her husband's relatives to be clubbed to death to ensure that the throne remains to her husband – **King Thibaw**.

44. **French Soldiers** arrived to fight the **Battle of Marne in World War I** – not on foot or by military airplane or military Vehicle – but by **Taxi Cabs**.

45. The 16th Century **Danish Astronomer Tycho Brahe** lost his **nose** in a duel with one of his students over a Mathematical Computation. He wore a silver replacement nose for the rest of his Life.

46. **"The Collosseum"** of Rome constructed between 70 AD & 80 AD received its name not for its size, but for a colossal statue of the Roman Emperor **"Nero"** that stood close by, placed there after the destruction of his palace.

47. The **Nobel Peace Prize** was first awarded in **1901** to **Jean Henry Dunant**, who was the founder of the **Swiss Red Cross**.

48. The **Aztec Indians** in Central America used animal **blood mixed with cement** as a mortar for their buildings, many of which still remain standing today!

49. **The Tower of London** for which construction began in **1078** by **William the Conqueror**, once housed a zoo. It also served as an observatory, a mint, a prison, a royal palace & at present, the home of the Crown jewels.

"The Collosseum" of Rome constructed between 70 AD & 80 AD received its name not for its size, but for a colossal statue of the Roman Emperor "Nero" that stood close by.

50. The **first dictionary of American English** was published on **April 14th, 1828** by American Author **Noah Webster**. It contained **70,000 words** in all.

51. The first time an **enormous amount of Clothing** was needed all at once was during the **American Civil War** (1861-1865), when the Union needed hundreds of thousands of **uniforms** for its troops. It was only after this that the **ready-made clothing industry** came into existence.

52. The Universally Popular **"Hersheys Chocolate Bar"** was used as a currency overseas during **World War II** as **currency**. Wow, how innovative & exciting it would have been to use chocolates as currencies!!

53. The only **South East Asian country** that has never been colonized by a **Western Power** is **Thailand**. Despite immense pressure from European powers, Thailand escaped colonial rule by maintaining strong rulers & exploiting the tension between colonizing powers - namely France & Great Britain - which had spheres of influence across neighboring countries in Asia.

54. **The right arm & torch** of the **Statue of Liberty** crossed the Atlantic Ocean three times. It first crossed for display at the 1876 Philadelphia Centennial Exposition & in New York, where money was raised for the foundation & pedestal. It was returned to Paris in 1882 to be re-united with the rest of the statue, which was then shipped back to U.S.

55. **Traffic engineering** was not developed in London, New York or Paris, but rather in **ancient Rome**. The Romans of course were noted road Builders. Best example is the "Appian Way" – which connected _Rome_ to _Brindisi_, _Apulia_, in southeast _Italy_ & was one of the first paved Roads. In Rome itself there were actually stop signs & even alternate side-of-the-street parking.

56. The **oldest Functioning Post Office** in the world is located in the village of **Sanquer**, located in the Scottish Lowlands. It has been operating since **1712**!!

57. When **Napoleon** wore **black silk handkerchiefs** around his neck during a battle, he always **won**. At **Waterloo**, he wore a **white cravat** & lost the battle & his kingdom.

58. **John Tyler**, the 10th U.S President had a **son** when he was **63**. That son fathered two sons of his own when he was 71 & 75. Those two sons are still alive. In other words, a man who became president in **1841** has still **two living grandchildren**!

59. **The Ottoman Empire** was one of the **longest lasting empires in history**. It spanned the years between **1299 & 1922** and through its duration covered much of south-eastern Europe and parts of northern Africa.

60. **Arnold Sommerfeld**, a theoretical physicist, was doctoral supervisor to 4 physicists who went on to win **Nobel Prizes** in physics, and academic supervisor to 2 others who also won. He himself was **nominated 81 times**. Despite of his hard work, he was never able to achieve the big prize, and hence, died without a Nobel Prize.

61. There were **57 countries** involved in **World War II**. Mainly, there were 2 groups - **"Allies"** consisting of the U.S joining hands with the U.K & **"Axis"** consisting mainly of Italy, Germany & Japan.

62. Since the **United Nations** was founded in **1945**, there have been **140 wars**! No wonder there has been a huge debate over the role of the U.N!

63. A Japanese soldier didn't know **WWII** had ended and hid in the jungle for **28 years**! **Shōichi Yokoi** was a sergeant in the Imperial Japanese **Army** during WWII, and in **1943** he was transferred to the island of Guam. On January 24th, 1972, 27 years after the end of the war, Yokoi was discovered by two local men who were out checking their shrimp traps. Yokoi, who thought the war was still going, saw them as a threat and attacked them. Luckily they were able to subdue him without major injury.

64. **Sunglasses** were invented by the **Chinese** in the **12th Century**. Early Sunglasses served a special purpose & it wasn't to block the rays of the sun. For Centuries, **judges** had routinely worn **smoke-colored quartz lenses** to conceal their eye expressions in Court. In 1929, **Sam Foster**, founder of the Foster Grant Company sold the first pair of Foster Grant sunglasses on the Boardwalk in Atlantic City, NJ.

SCIENCE

V. Science

1. World's **First Cloned Sheep's** name is **Dolly.** Dolly was cloned through a process known as **nuclear transfer** and was born on **5th July 1996** at the Roslin Institute in **Edinburgh**.

2. **Oppenheimer,** an **American Physicist** at the University of California, Berkeley is the **Father of Atomic Bomb** for his role during World War II. He also made important contributions to the theory of cosmic ray showers, and did work that eventually led toward descriptions of **quantum tunneling**. In the 1930s, he was the first to write papers suggesting the existence of what we today call **black holes**.

3. The **Smallest Bone in the human body** is the **Stirrup Bone** located in ear.

4. **Ants** do not sleep!! As per research, ants do sleep but not like humans; they usually sleep in cycles or patterns with each sleep lasting 8-10minutes!!

5. The **World's Smallest mammal** is the **Bumble-bee Bat** which weighs less than a penny! The Etruscan Shrew also shares this rare distinction. The Shrew weighs less than 2 grams!

6. **The Cornea** is the only **tissue** in the Human body that does not contain any **blood vessels**.

7. **William Harvey** an English Physician 1st discovered **Blood Circulation System** in **1667**. It is said that prior to this critical finding, people used to believe that lungs are responsible for the circulation of Blood!

8. **Enamel** is the **hardest substance** in the **Human body**.

9. The **Silk** that is produced by **Spiders** is said to be stronger than Steel based on density as Spider silk's density is 6 times lesser than steel of the same weight.

10. The **World's heaviest insects** are the **Goliath Beetles** from the family Scarabaeidea weighing **115 grams** and **11.5 cm in length**!

11. A **Chameleon's tongue** is as long as its body and head and can shoot out as fast as **16 feet/sec**!

12. The fastest moving bird is the **Peregrine Falcon** at a speed of around **350 km/hr**!

13. **Isaac Newton** was the first person to be **knighted** for Scientific achievement. He was knighted by **Queen Anne** in the year **1705**.

14. **Lachanophobia** is the fear of Vegetables!!

15. **Henry Becquerel**, a French Physicist discovered **"Spontaneous Radioactivity"** in **1896**, though fascinatingly he had been studying the field of Optics previously. It was because of his contact with the renowned physicist - Rontgen that began to take an active interest in Radioactivity.

16. Because of the **rotation of the earth**, an object can be thrown farther if it is thrown west

17. A **World record 328 Pounds Ovarian Cyst** was removed from a woman in Galveston, Texas in **1905**!! Post surgery her weight reduced from 708 pounds to 400 pounds!!

18. **"Natron" (Hydrated Soda ash)** was used in **Egyptian mummification** because it absorbs water and behaves as a drying agent.

19. An **ant** can survive for upto **two weeks underwater**! An ant can close the pores present on its bodies during the submergence in water and also lower necessary bodily functions. In this state, the ant required **20 times** less oxygen than it needs while sleeping. The lower the water temperature, the easier it is for the ant to maintain this lower metabolism and the longer it survives below water.

20. **Paleontology** is a branch of Science dealing with **Ancient Creatures**.

21. The only **two mammals** that can **see backwards** are **Rabbit & Parrot**. Rabbit eyes are placed high and to the sides of the skull, allowing the rabbit to see nearly 360 degrees, as well as far above their head.

22. **Pigs** are the **4th most intelligent** animals!! One of the examples of this fact is that they can trace their home even from a distant location.

23. A **Dragon-fly's** eyes contains **30,000 Lenses**! But, unfortunately they cannot see the details very well. In fact, a human eye has only one lens and sees better than a dragonfly.

24. **Kiwis** are the only Birds that hunt by **sense of smell**.

25. The **Human Bone** is 5 times stronger than **Steel** & 4 times stronger than **Concrete** although the **"Thigh Bone"** is 8 times stronger than Concrete!!

26. **The Science of Genetics** was founded by a German Scientist – **Gregor Mendel** who demonstrated inheritance in Pea plants in 1866.

27. **Platypus** & the Spiny Ant-eater are the only **mammals** which **lays eggs**! This is because they are of the order/class named "Monotremes", due to which they have a reptilian gait with legs on the sides rather than underneath the body and a single duct for urine, feces and sex instead of multiple openings.

28. **Silicon** is the **2nd most abundant element** in the **Earth's Crust.** That is due to the fact that Silicon is a key component of sand, which is found abundantly on earth.

29. **Selenology** is the astronomical study of the **Moon's** Geology, Physical Characteristics & Origin.

30. **Sun-spots** were first observed telescopically in 1610 by the **English Astronomer Thomas Harriot & Frisian Astronomers Johannes & David Fabricus**. Sunspots are temporary phenomena on the photosphere of the Sun, that appear visibly as dark spots compared to surrounding regions.

31. **The Ivory Billed Woodpecker** is the **rarest bird** in the world due to reports of its near extinction arising from the destruction of its Habitats in USA. This can be related to the fact that it was not seen since 50 years until spotted in Arkansas, USA in 2005!

32. The **longest Solar Eclipse** ever occurred on **20th June, 1955** at west of Phillipines lasting **7 minutes & 8 secs**.

33. **Ornithology** is the branch of Science dealing with the **study of birds**.

34. An **Octopus** has **three hearts**!! Two of the hearts work exclusively to move blood beyond the animal's gills, while the third keeps circulation flowing for the organs.

35. The **forests of Patagonian cypresses** (Fitzoya Supressedes) in the Andean Mountain ranges in Southern Chile & Argentina has an **average age of 2500 years** and are thus the **longest living forests** in the world!

36. **The Platypus** is one of the few **venomous mammals** and was first **encountered** by the Europeans in **1798**.

37. There is **0.2mg** of **Gold** in the **blood** of our body! Most of that gold is diffused with our blood. However, you had need to bleed 40,000 humans in order to collect enough blood to distill 8g of gold!

38. **Apollo 17** crew were the last men to land on the **moon**. It was overall the 6th landing on the moon from the US.

39. **Snails** have **14175 teeth** laid along 135 rows on their tongue!

40. The **Human body** has **78 organs**.

41. The **bird** with the **most feathers** is the **Whistling Swan Cygnus columbianus** which can have up to **25,000 feathers** during Winter.

42. **The first man made item** to exceed the **speed of Sound** is the Bull-whip or **Leather whip**. When the whip is snapped, the knotted end makes a "crack" or popping noise. It is actually causing a mini Sonic boom as it exceeds the speed of Sound.

43. The **first Space-craft** to visit the planet **Venus** was **Mariner 2** in 1962.

44. A **Cheetah** doesn't roar like a Lion, but **purrs** like a cat (Meow)!

45. The Planet **Saturn** has a **density** lower than water. If there was a bath-tub large enough to hold it, Saturn would float!!

46. **Bees** have **5 eyes**. There are 3 small eyes on the top of bee's head and 2 larger ones in front.

The Planet Saturn has a density lower than water. If there was a bath-tub large enough to hold it, Saturn would float.

47. There is no element on **Mendeleev's Periodic Table** of elements abbreviated, either partially or fully with the letter **J**. That's worth noticing!!

48. **Birds** cannot go into **Outer Space**, because they use **gravity** to assist them in Swallowing, so they had quickly choke & die in a non-gravity Environment.

49. Until going offline in March 2012, **Kashwazaki Karwa Nuclear Power Station** in Japan had a total Output of **8212 Mew** making it the **Largest Nuclear Power Station** in the World. It supplied electricity to 16 million House-holds & was the fourth largest Electric generating station in the World.

50. Certain **Fire-flies** emit light so **penetrating** that it can pass through wood & even Flesh!

51. The **smallest eggs** laid by a bird are of **West Indian Vervian Humming bird** of **10mm length** & **0.375 grams** in weight!!

52. **Females** have **500 more genes** than **males**, and because of this are protected from things like **color blindness** & **Haemophilia**.

53. **Cows** are the only **mammals** that **pee backwards**!

54. **Dolphins** can kill **Sharks** by ramming them with their **Snout**. The Snout is made of very strong and thick bone, and has a hard rounded end which is enough to kill the shark!

55. **Humans** shrink in size after the age of **30**. Generally a man shrinks about **3 cm** in his lifetime and a **woman about 5 cm**. This is part of the natural ageing process as the body loses muscle and fat.

56. **Flamingoes** live remarkably long lives of upto **80 years**!

57. It takes an interaction of **72 different muscles** to produce **Human Speech**! That's why we say that god has created humans in a very systematic & beautiful manner.

Flamingoes live remarkably long lives of upto 80 years!

58. It takes a **sloth** upto **six days** to **digest** the food it eats!

59. Female **Alligators** lay about **40 eggs** that hatch in **60-70 days**.

60. The **slowest growing finger nail** is on the **thumb nail** and the **fastest growing is the finger nail** on the **middle finger**.

61. The Storage Capacity of the **Human Brain** exceeds **4 Terrabytes**!! Not only that the human brain can recognize 10,000 different odors and 7 million colors! How astonishing is that when you see the numbers!!

62. **Koala** - an **arboreal herbivorous marsupial** of **Australia** almost **never drink water**. They get fluids from the **eucalyptus leaves** they eat!

63. **Lady Peseshet** was known to be the **World's first known female physician**. She practiced during the time of the pyramids, which was the **fourth dynasty**.

64. There are **Laser Reflector mirrors** on the **moon**. **Astronauts** left them so that laser beams could be bounced off of them from Earth. These beams help give us the approx. distance to the moon.

SPORTS

VI. Sports

1. Former English Footballer - **Gary Linekar** is the only one to have won **3 consecutive** Golden Boot Awards with 3 different Clubs. The 3 nos. of clubs are Leicester City, Everton & Tottenham Hotspur.

2. **Graeme Smith** is the **most successful Test Captain** in the history of Cricket with **53 Test wins, captaining 109 matches** till his retirement in 2014.

3. **Australia** achieved the largest victory ever in an International **Soccer match** when they beat **American Samoa 32-0 in 2001**!

4. From **2010** onwards, there are 17 Professional **Boxing** Weight classes.

5. Before 1850, **Golf balls** were made of **leather** and were **stuffed with feathers**.

6. **India** won its **1ˢᵗ Gold medal** at the Amsterdam Olympics in **1928** for Hockey. India's performance earned them rave reviews and while only three persons saw them off on their journey to London, massive crowds thronged the Bombay port to welcome the new Olympic champions!!

7. Former American Basket-ball Champ, **Bill Russell** is the player with the **most no. of NBA Basketball Championships**. He played for the **Boston Celtics**.

8. The biggest win in **Hockey** was when **Canada** beat **Denmark 47-0** in 1949 Olympics in Sweden!!

9. The 1ˢᵗ ever **ODI Cricket** match was played between England & Australia at Melbourne in 1971. When the first three days of the Third test were washed out, officials decided to abandon the match and instead play a one-off one day game consisting of 40 eight-ball overs per side. Australia won the game by 5 wickets!!

10. In Cycling in **Tour de France**, *"The King of Mountains"* signifies the **Best Climber in the race** and he wears a White Jersey with red dots.

11. Former Indian Cricketer **Rahul Dravid** has faced the **most no. of Balls in Test cricket** with **31,258!!** He also has a Record for grabbing the **most no. of catches** by a Fielder in **Test Matches – 210** in 164 matches!

12. The **most runs scored** by a side on any day on a **Test match is 509** by **Srilanka against Bangladesh** at Colombo in 2002 (starting the day with 32 & ending at 541/9).

13. **Miroslav Klose, Germany** has scored **highest no. of goals(16 nos.)** overall in **FIFA World Cups**. He retired from Germany's national team on 11ᵗʰ August, 2014.

14. **Cricket** has been included in only **1 Olympic Games Event in 1900 at Paris** where just 2 teams took part (Great Britain & France) & in which GB won by 158 runs!!

15. The **NBA's** (National Basketball Association) leading points scorer is former American Basketball Player **Kareem Abdul Jabbar** who scored **38,387 points**. He played between 1969-1989.

16. **American Swimmer, Micheal Phelps** has the record for winning the most no. of Olympic medals ever. He has won **22 Medals** (18 Gold, 2 Silver & 2 Bronze) In April 2014, Phelps announced that he would come out of retirement, and would enter an event later that month

17. The **most economical figures ever in an ODI** was 10overs 3 runs for 4 wickets by medium pacer **Phil Simmons** of West Indies against Pakistan in 1992.

18. Since **1896**, the beginning of the modern Olympics, only **Greece & Australia** have participated in every Games.

19. The official **Olympic Flag** was created in 1914 by French Historian & Educator **Pierre de Coubertin** and was first flown during the 1920 Olympic Games.

20. Former English Footballer **Alan Shearer** has the most no. of **Goals (260 nos.)** in the English Premier League.

21. **David Houghton** of Zimbabwe was a wicketkeeper batsman cum Coach for the Cricket team as well as the **Goal-keeper for the National Hockey team**! That's a super ambitious sportsman!!

22. **Kite-flying** is a Professional Sport in Thailand. Basically, a kite flying competition is organized in the month of April, which goes on for 15 days with two teams with traditional names as "Chula" & "Pakpao" participating.

23. The **1st Olympic Race** held in 776 B.C was won by "Corus" a **chef**!

24. Pakistani Cricketer **Shahid Afridi** has hit the highest no. of Sixes **(351 nos.)** in 398 ODIs till 20th March 2015.

25. **France** has hosted the **1st Winter Olympic** Games in 1924.

26. Former West-Indian great **Vivian Richards** also played **football** for Antigua in 1974 apart from being one of the greatest cricketers of all times. He also was the only West-Indian Captain never to lose a Test series.

27. **Rocky Marciano** was an **American-Italian Boxer & World Heavyweight Champion** from 1952-56. He was the only Boxer to go undefeated in his career of **49 wins** in which 43 of them were knockouts!

28. **London** is the only city to have hosted the **Olympic Games** on 3 different occasions- **1908, 1948 & 2012.**

29. **The game of Rugby** was originated at Rugby school in 1823 when a Pupil – **William Webb Ellis** picked up a Foot-ball and ran away with it.

30. **Ajit Agarkar of India, Bob Rolland of Australia & Mohammed Asif of Pakistan** hold the dubious joint record for the most consecutive no. of ducks – **5 nos.** in International Cricket.

31. The word **"Gymnasium"** comes from the Greek root "Gymnos" meaning nude; the literal meaning of Gymnasium is **"school for naked exercises"**. Athletes in the ancient Olympic Games would participate in the nude!

32. **Vasaloppet, Sweden** is the oldest, longest and the biggest **Cross-country Ski race** in the world (90kms). Every year around 30,000 people compete in the race.

33. The **longest ever Tennis match** in history took place at Wimbledon 2010 when **John Isner** of the United States beat **Nicolas Mahut** of France 6-4,3-6,6-7(7),7-6(3), 70-68 in a match that lasted 11 hours & 5 minutes played over 3 days – June 22,23,24!

34. The **1912 Olympics** held at Stockholm, Sweden was the last time when **Gold medals** were made of pure gold. Since then they have been Silver with Gold Plating.

35. American **NASCAR Driver Richard Petty** holds the record for the most no. of wins – **200 wins & 7 championships**. He was inducted into the inaugural class of the NASCAR Hall of Fame in 2010.

Former Indian Cricketer Rahul Dravid has faced the most no. of Balls in Test cricket with 31,258!! He also has a Record for grabbing the most no. of catches by a Fielder in Test Matches – 210 in 164 matches

36. ***Sachin Tendulkar, Waqar Younis & Salil Ankola*** *all made their debuts in the same test at Karachi in 1989.*

37. *It is customary for the **Jockeys** to be paid in Coins, no matter how large their winnings are! The fees are between $30 per mount and $100 per mount depending on the purse structure at each particular track. Those are the fees jockeys receive if they do not finish first, second or third.*

38. *In ancient Greek, **Wrestling** matches were in the **nude** and the match did not end until one of the competitors got aroused! How embarrassing that would have been, had the rules remain the same!!*

39. *Minimum no. of Seats for a **stadium** to attain UEFA 5 star ratings is **50,000**. The Camp Nou stadium located in city of Catalan, Spain is the largest stadium in Europe having a capacity of 98,000!!*

40. *In 1997, 16-year-old Swiss Sportswoman, **Martina Hingis** became the **youngest women's tennis player** to be **ranked no. 1** in the world since the rankings began in 1975.*

41. *In 2002, **Germany's Oliver Kahn** became the **first goalkeeper** to win the **Most Valuable Player** of the World Cup.*

42. *Former Australian Wicket-keeper Batsman **Adam Gilchrist of Australia** has the most no. of dismissals **(417)** in One day Internationals.*

43. ***Boxing*** *is the only Olympic sport where wearing a **Beard is prohibited.** In my opinion, this rule should be exempted for the promising Muslims & Sikh boxer professionals who cannot shave their beard and want to continue boxing.*

44. ***Bangladesh*** *is the **most populous Country** in the world to have never won an **Olympic Gold Medal**.*

45. ***Sculls & Egg-beaters*** *are two of the basic **skills or Hand Movements** in **Synchronized Swimming** to give stability & height in performing Strokes.*

46. Czech Tennis Star **Martina Navratilova** was the **first** to win the **women's singles tennis title at Wimbledon nine times.**

47. **Wayne Douglas Gretzky** is a **Canadian** former professional **ice hockey player** and former head coach. He played 20 seasons in the **National Hockey League (NHL)** for four teams from 1979 to 1999. He is the **leading point-scorer in NHL history**, with more assists than any other player has points, and is the **only NHL player to total over 200 points in one season** – a feat he accomplished four times!!

48. The **Tour De France** Race is **2300 miles** long! The modern editions of the Tour de France consist of 21 day long segments (stages) over a 23-day period.

49. Indian Cricketer **Virat Kohli** has a flurry of records & has achieved a lot at a young age of 26. Some of his records include – The **fastest to reach 25 ODI Hundreds, the fastest Indian Centurion** in the world when he smashed a **100* (52)** against Australia at Jaipur. He also is the fastest in the world to score 7000 ODI runs.

50. In 1938, American Tennis Champ – The Late **Donald Budge** (1915-2000) was the first Person to win all the **four Grand Slams** in a single year!

51. In 1870, **British Boxing Champ Jim Mace** & **American Boxer Joe Cuburn** fought for **three hours & 48 minutes** without landing a single **Punch**!

52. **Nagoya Grand Bowl** is now the **biggest bowling alley** in Japan. It has **156 lanes** on three floors with 52 lanes each. One floor is synthetic and the others are wood. Formerly, the alley had even 260 lanes in two buildings. One of the buildings was demolished and the space is now part of the Freeway Interchange.

The Tour De France Race is 2300 miles long!

53. The **Motto for the Olympic games** is **"Citius Altius Fortuis"** if translated it means **Faster Higher & Stronger**. The motto was proposed by French Historian Pierre de Coubertin on the creation of the **International Olympic Committee** in **1894**.

54. **Ra Kyung-min (South Korea)** needed just **six minutes** to beat **Julia Mann (UK)** 11-2, 11-1 during the **1996 Uber cup in Hong Kong on May 19 1996**, a record for the **shortest competitive badminton match**!

55. **Judo**, "the gentle way" is the **first martial art** accepted at the **Olympic games**. It was first included in the Summer Olympic games at the **1964 games, Tokyo**.

56. The **game of Tennis** was invented in the **12th century** by **French Monks**. It was not until the 16th century that rackets came into use. It was popular in England and France, although the game was only played indoors where the ball could be hit off the wall. Henry VIII of England was a big fan of this game, which historians now refer to as real tennis.

57. In **October of 1998** a bolt of **lightning** killed an entire **11-man soccer team** from the **Democratic Republic of Congo in a match played at** Kasai Province in Congo. The opposing team was completely unharmed. Thirty more people had burn injuries.

58. Former Sri-lankan Test player **Duleep Mendis** is the **only batsman** to score **identical twin hundreds** in Test Cricket. He scored **105** in each of his innings which helped to draw the match against **India in 1982**.

59. **Laura Trott**, an English Track Cyclist who was born with a collapsed Lung & diagnosed with Asthma is now a **double Olympic Gold Medalist in track Cycling**. Indeed an inspiration for all those little children aspiring to be an athlete, but suffering from Bronchial diseases!!

60. **Tiger Woods** is the first athlete to have been named **"Sportsman of the Year"** by magazine **Sports Illustrated** two times in **1996 & 1997**.

61. *Etymologically, the word* **"coach"** *is a derivative of the kind of vehicle, which comes from the Hungarian word* '**kocsi**'. *This word means "vehicle" – a horse-drawn carriage (in English – coach). The word* '**kocsi**' *comes from the name of the village in* **Hungary** *where it was first built.*

62. *Former Golf Professional* **Arnold Palmer** *became the* **first player** *to cross* **$1 million** *as Career* **PGA (Professional Golfer's Association)** *Earnings in* **1968.**

63. **William Joseph Klemm** *was known as* **"Father of Baseball Umpires".** *He worked in a record* **18 World Series.** *He has thus served the* **most no. of games** *served –* **5368.**

64. *During* **World War II**, *the National Football League (Rugby) faced a crisis unimaginable today: a shortage of players. By* **1943**, *so many players were in the armed forces that the league was forced to fold one team (the Cleveland Rams) and merge two others: the* **Pittsburgh Steelers** *and the* **Philadelphia Eagles**. *Thus were the* **"Steagles"** *born. The Steagles included military draft rejects, a superstar lured out of retirement, and even a few active-duty servicemen who got leave for the games. Yet, somehow, this motley crew posted a winning record-the first in Eagles' history and the second in Steelers' history!*

65. *Former* **English** *Opening Batsman* **Geoffrey Boycott** *of England was the* **first player to face a ball in ODI Cricket** *in the first ODI Match played between* **Australia vs England** *at MCG, Melbourne in 1971. Former Australian Pacer* **Graham Mc-Kenzie** *was the bowler.*

66. **Late Colombian Player Andres Escobar** *was murdered after* **Colombia's** *loss to U.S when he accidently scored an own goal which gave the Americans a* **1-0 lead** *to knock out the Colombians out of World Cup. After Escobar returned home to Colombia, he was* **gunned** *down three times on* **2nd July 1994**. *Some people get so desperate about their country winning that they forget that after all we are humans!!*

67. ***The second day of the Lord's test match*** between **England and West Indies** in 2000 is the only time in the history of test cricket that a part of **all 4 innings** have been played on the **same day**! WI were **267/9** at the end of day 1. They faced just one delivery on day 2 and were all out for 267. England made 134 and WI were all out for 54, both happened on day 2. England came out to bat, faced 1.1 overs and were 0/0 at the end of day 2.

68. Former Zimbabwean player – **Tatenda Taibu** is the **youngest Test Captain** at the age of **20 years & 358 days**, while Former Bangladesh Player **Rajin Saleh** became the **youngest ODI Captain** at the age of **20 years & 297 days**.

WORLD RECORDS

VII. WORLD RECORDS

1. *The World record for the most number of days without sleep is 264 hours (11 days) by **Randy Gardner of USA in 1965**. It is said that a human being can stay without eating for 11 days at a stretch!!*

2. ***Tom Sietas, a German Diver** holds the record for holding breath underwater for **22 minutes & 22 seconds**!!*

3. *The Current Pi champion is **Hiroyuki Gotu of Japan** who memorized **42,195** places of Pi! That's a living Super-computer amongst us!!*

4. ***Bernard Clemmens** of London managed to sustain a **fart** for an officially recorded time of **2 minutes 42 seconds**!! What a control & stamina!!*

5. ***Chandra Bahadur Dangi** of Nepal holds the World record for the **shortest man on earth** with a height of **54.6cms (21 inches)**! Unfortunately, he died on 4th September 2015 due to pneumonia.*

6. *The World record for the **longest Radio show broadcasted** is 73 hours by **Noria Nieski & Tolga Alka** for KISS FM Radio in Germany in 2011.*

7. ***Scottish Woman Connie Dennison** holds the World Record for the **oldest Yoga teacher** at the age of **99**! She hopes to continue teaching even after turning 100 later this year!!*

8. ***Nabil Karam of Lebanon** holds the World Record for the **largest collection of Model Cars – 27,777**!!*

9. **The Tokyo Sky-tree tower**, Japan holds the World Record for the **tallest "Tower" in the world** with its height measuring upto **634 mtrs**.

10. **USA** has the most no. of **Nobel Peace Prize** Recipients – **272 nos.** with the latest being US President Barack Obama in 2009.

11. The **largest Anamorphic (3D) Painting** measures 106.3 mtrs and was created by **UK Painter Joe Hill** at West Indian Quay, **London** on 17 november 2011.

12. **Colin Furze of UK** has super-charged a **Mobility scooter** so that it can reach a top speed of **115km/hr**. It took him 3 months to do so featuring 5 gears, a 125 cc motor-bike engine and twin exhausts. What an effort!!

13. The **farthest distance** ever travelled from Earth by humans is **400,171 kms** by the crew of **Apollo 13** on **15th April,1970**.

14. The **greatest weight lifted** with a human Tongue is **12.5 kg** by **Thomas Blackthorne of UK** on the sets of El Olympico, Mexico on 1st August 2008.

15. **American Architect – Bryan Berg** broke his own Guinness World Record of constructing the largest Structure by **stacking cards**! He created a replica of the **Venetian Macau** using **218,792 cards** & it took him **45 days** to complete it!! That's creativity & patience at its best!! What an inspiration!!

16. The **tallest living person** on earth is **Sultan Kosen** from Ankara, Turkey measuring **8 feets 3 inches (251 cms)** on 8th Feb 2011. He married a Syrian woman in October 2013.

17. **Expedition 1's Soyuz-TM** was launched to the International Space Station on 31st October 2000 and its **crew of three** remained in space for **136 days** which is the world record for the **longest uninterrupted Human presence in space** with more than 10years of continuous occupation at the International Space Station.

18. The World record for the **Smallest Road-worthy car** is the one created by **Austin Coulson** of USA. Measured in Texas, it is **63.5cm high** (25 inches), **65.41cm wide** (2 ft 1.75 inches) and **126.47cm long**. The vehicle is licensed to be driven on public roads with a speed limit of **40 km/hr (25mph)**!

19. The World Record for the **Oldest living Person** is currently held by **Susannah Mushatt Jones,** an American Supercentenarian who is aged 116 years 75 days. She is partially blind & deaf & is confined to her wheelchair.

20. The World Record for the most no. of Children born to one Mother is **69** in the case of the wife of a Russian Peasant – **Feodor Vassilyev**!!

21. The World Record for the **most Expensive Wedding** was the wedding of **Vanisha Mittal, daughter of Lakshmi Mittal with Amit Bhatia**, an investment banker. The six day affair was held at Versailles, France costing the bride's father a bill of **$ 55 Million**!! Lakshmi Mittal is a Indian steel tycoon residing in UK.

22. The **World record** for the **tallest living dog** in the world is **"Zeus" (US), the Great Dane** who measured **1.118 m** owned by Dennis Doorlag from Otsego, Michigan, US.

23. The World Record for the **Largest Collection of Barbie dolls** is of **Bettina Dorfman** of Germany who has **15,000** different Barbie dolls. She has been into it since 1993.

24. The **largest Drum-set** is owned by **Dr Mark Temperato of USA** and is comprised of **340 pieces**!! He is a member of the Professional Band **"Jesus the Soul Solution".**

25. The World Record for the **most no. of people on a two-wheeler** is **54** and was achieved by the **Army Services Corps. Motorcycle Display team – "Tornadoes" in Bangalore, India**. The men rode a 500cc Royal Enfield at a distance of **1,100 mtrs**!! That's what we called a perfect example of some superb Team-work & Coordination.

26. *The World Record for **the longest hair** is by **Xle Oluping** from China whose hair measured **5.27 metres** in May 2004!! The **Fastest to 1 million followers** on Twitter is Hollywood actor **Charlie Sheen** in **25 hours 17 minutes** on 1st & 2nd march 2011.*

27. *The **first to reach 1 million followers** is actor **Ashton Kutcher**. But Kutcher might want to keep a closer eye on his account, which was hacked in 2009.*

28. ***Hōshi ryokan** (Japanese traditional inn) in the area of Komatsu, Japan is the **World's oldest Continuously operating Hotel** which was founded in the year **717**, according to the Guinness World Records and the world's oldest continuously operating company after the liquidation of Kongō Gumi in 2006. Hats off to the people who are able to salvage the heritage of the Hotel.*

29. ***Terry Burrows (UK)** cleaned **three standard 114.3 x 114.3 cm** (45 x 45 in) office windows set in a frame with a 300 mm (11.75 in) long squeegee and 9 litres (2 gal; US 2.37 gal) of water in **9.14 sec** at the **National Window Cleaning Competition** in Blackpool, UK, on 9 October 2009 which is a World Record!*

30. ***The most Christmas trees chopped** in **two minutes** is **27** by **Erin Lavoie (USA)** achieved on the set of Guinness World Records in Germany, on 19 December 2008. Now that's ridiculous deforestation (on a lighter note)!!*

31. ***Kanchana Ketkaew (Thailand)**, lived in a glass room measuring **12 m²** (130 ft²) containing **5,320 scorpions** for **33 days and nights** at the Royal Garden Plaza, Pattaya, Thailand, from 22 December 2008 to 24 January 2009. Over the 33 days, she was **stung thirteen times!!***

32. ***The longest time survived trapped underground** is **69 days** by "The 33 of San Jose", **(32 Chilean and 1 Bolivian)**, who were trapped 688 m (2,257 ft) below the surface after the collapse of the San José copper-gold mine, near Copiapó, Chile, on 5 August 2010. That's really a miraculous escape!!*

Knowledge Capsule

33. ***The most expensive pizza***, *commercially available, is a thin-crust, wood fire-baked pizza topped with onion puree, white truffle paste, fontina cheese, baby mozzarella, pancetta, cep mushrooms, freshly picked wild mizuna lettuce and garnished with fresh shavings of a rare Italian white truffle, itself worth **£1,400 (then $2,500) per 1 kg (2 lb 3 oz)** served at the **Maze Restaurant, London**.*

34. ***The most needles inserted*** *on the head is **2,009** and was achieved by **Wei Shengchu (China)** on the set of Lo Show dei Record, in Milan, Italy, on 11 April 2009.*

35. *Comprising 242 individual frames with a frame size of only 45 nanometres by 25 nanometres (45 x 25 billionths of a metre), **the smallest stop-motion film** was created by **IBM Research Laboratories** in San Jose, California, USA, from **29 January to 6 February 2013**. Individual molecules of carbon monoxide were "placed" as pixels on a copper sheet to create each frame of the film.*

36. *The record for the **most one arm push-ups** completed in **one hour is 1,868** and was set by **Paddy Doyle (UK)** at the Munster Arms Hotel, Sparkbrook, UK, on 27 November 1993.*

37. ***Kevin Fast (Canada)*** *holds the World Record for **pulling the Heaviest Aircraft** – A **CC-177 Globemaster III**, weighing **188.83 tonnes** (416,299 lb), a distance of **8.8 m** (28 ft 10.46 in) at Canadian Forces Base in Trenton, Ontario, Canada, on 17 September 2009. The fastest time to solve a **Rubik's Cube is 6.24 seconds** and was achieved by **Feliks Zemdegs (Australia)** in the Kubaroo Open 2011 held in Melbourne, Australia, on 7 May 2011. Well he must be an android or a Super-Computer!!*

38. *The **longest street** in the world is **Yonge Street**, which starts in **Toronto**, on the north shore of Lake Ontario, and winds its way north then west to end at the Ontario-Manitoba-Minnesota border, which is over **1896 kms** long!! But surely, there were controversies surrounding this claim as a part of this exceptionally long street also includes part of Highways*

39. ***Abhijeet Baruah****, a 22 year old Indian Constable ran **Bare-feet** for **156.2 kms** at Jorhat, Assam on 30-31 January 2012 making it a World Record!!*

40. *The World Record Holder for the **Longest Domestic Cat** is held by **"Mymains Stewart Gilligan"** which is **123cms** long & is owned by Robert Hendrickson of USA!*

41. *The **Tallest Domestic Cat** is "**Savannah Islands Trouble**" which is **48.3 cms tall**, owned by Deby Maraspini of USA & was measured at the Silver Cats Cat show at Nevada, USA on 30th October 2011.*

42. ***Betty White*** *earned the record for "**Longest TV Career for an Entertainer (Female),**" spanning **74 years** of work in the industry. Making her debut in 1939, the queen of the small screen has appeared in some of television's most popular shows, including "The Golden Girls," "The Mary Tyler Moore Show," and "The Carol Burnett Show. Currently starring in "Hot in Cleveland," the legendary 91-year-old entertainer shows no signs of stopping!!*

43. *Westwood Studio's Computer game –* ***"Command & Conquer"*** *is the most* ***successful war game series*** *of all time according to* ***Guiness Book of World Records****. Well, I had an opportunity to play one of the games in this series - Red alert and I found it exceptionally designed.*

American Architect – Bryan Berg broke his own Guinness World Record of constructing the largest Structure by stacking cards! He created a replica of the Venetian Macau.

44. *As if fire-eating wasn't enough,* **Carisa Hendrix** *of* **Canada** *spiced up her attempt further by teething – gripping the torch by her teeth without taking a breath, fire side down. She practiced for a month for her* **2 minutes 51 seconds** *feat on "Lo Show dei Record" in* **Rome, Italy on 12th April 2012**. *She was another multi-talented individuals expertising in the fields of acting, dancing as well as teaching!!*

45. **The first expedition** *to reach the geographic* **South Pole** *was led by the Norwegian explorer* **Roald Amundsen**. *He and four others arrived at the pole on* **14 December 1911**, *five weeks ahead of a British party led by Robert Falcon Scott as part of the Terra Nova Expedition. Amundsen and his team returned safely to their base, and later learned that Scott and his four companions had died on their return journey.*

46. **Tony Collins (UK)** *spent* **77 hr 30 min** *on a* **hospital trolley** *in a corridor at* **Princess Margaret Hospital**, *Swindon, Wiltshire, UK, from 24-27 February 2001 before he was admitted to a bed. A diabetic, he was suffering with a virus.*

47. *The* **world record** *for* **rocking non-stop** *in a* **rocking chair** *is* **480 hours** *held by* **Dennis Easterling** *of Atlanta, Georgia.*

48. **Tom Owen** *of USA had* **nine Pick-up trucks** *each weighing* **3000 kg** *run over his* **stomach** *on the set of "La Show Dei Record" at Milan, Italy in* **26th April, 2009**!! *Here we have a live Superman with some extra-ordinary abilities!*

Tokyo sky-tree tower

49. ***The smallest waist*** belongs to **Cathie Jung** (USA, b. 1937), who stands at 1.72 m (5 ft 8 in) and has a corseted waist measuring **38.1 cm (15 in)**. Uncorseted, it measures 53.34 cm (21 in). Not sure whether she eats anything or not!!?

50. ***101 year*** old **Anthony Mancinelli** started cutting hair at age of 12 & he continues to work today – setting the world record for the **Oldest Barber**!

Roald Amundsen

MIXED BAG

VIII. Mixed bag

1. *Who played the Title Role in **Robin Hood**?*
   i) **Kevin Costner**
   ii) *Russell Crowe*
   iii) *Paul Newman*
   iv) *Mark Wahlberg*

2. *Which of these isn't an **event** in **Men's Gymnastics**?*
   i) *Horse Pummeling*
   ii) **Rope Climbing**
   iii) *Steady Rings*
   iv) *Parallel Bars*

3. *Which of these is a **European Wind**?*
   i) *Afghani*
   ii) **Mistral**
   iii) *Monsoon*
   iv) *Derivatives*

4. *Which **element** is named after **Scandinavian God of Storms & War**?*
   i) *Titanium*
   ii) *Sellenium*
   iii) *Zirconium*
   iv) **Thorium**

5. Which gas is used to fill **Tires** of **Racing cars**?
   i) **Nitrogen**
   ii) Hydrogen
   iii) Hydrogen Peroxide
   iv) Carbon Dioxide

6. Who is the **Lead singer** of Pop Band **"Greenday"**?
   i) Tommy Lee
   ii) Jack Patterson
   iii) Michael Tobowsky
   iv) **Billy Joe Armstrong**

7. Which country is called the **"Pearl of the Indian Ocean"**?
   i) Maldives
   ii) Sumatra
   iii) **Sri Lanka**
   iv) Thailand

8. **"Volvo"** brand of Automobiles is from?
   i) Germany
   ii) **Sweden**
   iii) France
   iv) Italy

9. Where will be the next **Summer Olympics** be held in 2016?
   i) Munich
   ii) Rome
   iii) **Rio-de-Janeiro**
   iv) Mexico

10. What was **New-york** called before 1664?
    i) **New Amsterdam**
    ii) New Berlin
    iii) New London
    iv) New Piccadilly

11. **"Ravioli"** *dish is from:*
    i) *Greece*
    ii) *Japan*
    iii) *Switzerland*
    iv) **Italy**

12. *Which out of the following teams has won the* **FIFA WC** *most no. of times?*
    i) *France*
    ii) *Netherlands*
    iii) **Uruguay**
    iv) *Paraguay*

13. *What is the* **average Life span** *of a* **Red Blood Cell**?
    i) *90 days*
    ii) **120 days**
    iii) *145 days*
    iv) *110 days*

14. *Where was* **"Solid Tea Blocks"** *used as a currency in* **1930's**?
    i) **Siberia**
    ii) *Iran*
    iii) *Netherlands*
    iv *Middle east*

15. **The Faroe Islands** *are the territory of which country?*
    i) *Norway*
    ii) *Spain*
    iii) *UK*
    iv) **Denmark**

16. *Who is Manufacturer of* **Formula one Safety Car**?
    i) **Mercedes**
    ii) *Audi*
    iii) *Porsche*
    iv) *BMW*

17. Which of the following is an ingredient of **Margarita Cocktail**?
    i) **Lime Juice**
    ii) Rum
    iii) Orange Juice
    iv) Raspberry Juice

18. Which **Formula one** racer lost the 2007 World Championship by 1 point, but won the 2008 one?
    i) Michael Schumacher
    ii) Kimi Raekonnen
    iii) **Lewis Hamilton**
    iv) Fernando Alonso

19. What kind of animal is a **Curlew**?
    i) Rodent
    ii) Fish
    iii) **Bird**
    iv) Reptile

20. Who said *"Death is the greatest of all Human Blessings"*?
    i) Plato
    ii) **Socrates**
    iii) Archimedes
    iv) Aryabhatta

21. Which of these actors won **best actor academy award** for two consecutive years?
    i) Clint Eastwood
    ii) Robert De-niro
    iii) **Tom Hanks**
    iv) Harrison Ford

22. Which country is also known as *"Suomi"*?
    i) France
    ii) Italy
    iii) **Finland**
    iv) Norway

23. *Who appears on a **100 $** note?*
    i) **Benjamin Franklin**
    ii) *George Washington*
    iii) *Franklin Roosevelt*
    iv) *Abraham Lincoln*

24. *In which team did English Footballer **David Beckham** play from 1993-2003?*
    i) **Manchester United**
    ii) *Arsenal*
    iii) *FC Internationalze*
    iv) *Bayern Munich*

25. *Which of these **birds** has the **Longest Nesting times** amongst these:*
    i) *Mallard*
    ii) **Hawk**
    iii) *Starling*
    iv) *Eagle*

26. *Founder of **Quantum Theory**?*
    i) *Ernest Rutherford*
    ii) **Max Planck**
    iii) *Albert Einstein*
    iv) *Frank Webber*

27. *In which country was the **Rubik's Cube** invented in 1974?*
    i) **Hungary**
    ii) *Japan*
    iii) *USA*
    iv) *Romania*

28. *Capital of **Vietnam**?*
    i) *Ho chi minh City*
    ii) **Hanoi**
    iii) *Vung tauiv*
    iv) *Phnom-Penh*

29. Which of these is a Japanese **Martial art**?
    i)   Taekwondo
    ii) **Kendo**
    iii) Kung-fu
    iv) Thangta

30. Which out of the following **trees** blooms the earliest?
    i)   Pear
    ii) **Almond**
    iii) Apple
    iv) Peach

31. Who takes the **Hippocratic Oath**?
    i)   **Doctors**
    ii) Lawyers
    iii) Scientists
    iv) Commandoes

32. Which of the following is not recognized as an **Olympic Sport** anymore?
    i)   **Softball**
    ii) Judo
    iii) Taekwondo
    iv) Beach Volleyball

33. Which City was previously called as "Salisbury"?
    i)   London
    ii) **Harare**
    iii) Milan
    iv) Oslo

34. Who in 1944 got elected as **US's president** for record 4th term?
    i)   F. Truman
    ii) **F. Roosevelt**
    iii) John Kennedy
    iv) Ronald Reagan

35. Which animal's **Pupil** isn't round?
    i) **Goat**
    ii) Owl
    iii) Lion
    iv) Wolf

36. Which of the following are breeds of **Mountain Rescue dogs**?
    i) Terrier
    ii) German Shepherd
    iii) **Saint Bernard**
    iv) Labrador Retriever

37. Which of the **Cocktails** is made from **Vodka**?
    i) **Screwdriver**
    ii) Tom Collins
    iii) Margarita
    iv) Admiral

38. What is protected by **Scrum Cap**?
    i) Elbow
    ii) Chest
    iii) **Ear**
    iv) Ankle

39. What is **Altocumulus**?
    i) Seaweed
    ii) **Cloud**
    iii) Stinging Nettle
    iv) Rock

40. Oldest **Architectural Style** amongst the following:
    i) Gothic
    ii) **Romanesque**
    iii) Baroque
    iv) Renaissance

41. Which of the following Countries **Currency** is *"Manat"*?
    i) Bolivia
    ii) **Azerbaijan**
    iii) Poland
    iv) Uzbekistan

42. Which of the following **elements** has the most **Stable** Electronic Structure?
    i) Helium
    ii) Strontium
    iii) **Neon**
    iv) Chromium

43. Which of the following **Cats** can swim well?
    i) Leopard
    ii) Lynx
    iii) Cheetah
    iv) **Tiger**

44. Which one of the following was NOT amongst the Seven Wonders of the World?
    i) **Iron Pillar by Dhava**
    ii) Hanging Gardens of Babylon
    iii) Light-house of Alexandria
    iv) Zeus Statue

45. Main **Religion** of **Japan**
    i) **Shintoism**
    ii) Buddhism
    iii) Taoism
    iv) Confucionism

46. Who was the longest serving **British Prime Minister** of the 20th Century?
    i) **Margaret Thatcher**
    ii) Winston Churchill
    iii) Tony Blair
    iv) Woodrow Wilson

47. What is an **Armageddon**?
    i) Spanish Navy
    ii) A group of European nobles
    iii) A South-American film
    iv) **A Destructive final battle that takes place in the Bible**

48. Which is the only **Country** in the world which has **every climate** in the world?
    i) Australia
    ii) **New Zealand**
    iii) Greece
    iv) Luxembourg

49. How did **Queen Cleopatra** die?
    i) Heart attack
    ii) **She forced a Wasp to bite her**
    iii) She was murdered by her Husband
    iv) Eaten by a Tiger

50. Which Hollywood star has made it to cover of **Life** maximum no. of times?
    i) Lady gaga
    ii) Marilyn Monroe
    iii) **Elizabeth Taylor**
    iv) Richard Burton(Answers to be followed on the next page...

IX. Answers

1. *i)*
2. *ii)*
3. *ii)*
4. *iv)*
5. *i)*
6. *iv)*
7. *iii)*
8. *ii)*
9. *iii)*
10. *i)*
11. *iv)*
12. *iii)*
13. *ii)*
14. *i)*
15. *iv)*
16. *i)*
17. *i)*
18. *iii)*
19. *iii)*
20. *ii)*
21. *iii)*
22. *iii)*
23. *i)*
24. *i)*
25. *ii)*
26. *ii)*
27. *i)*
28. *ii)*
29. *ii)*
30. *ii)*
31. *i)*
32. *i)*
33. *ii)*
34. *ii)*
35. *i)*
36. *iii)*
37. *i)*
38. *iii)*
39. *ii)*
40. *ii)*
41. *ii)*
42. *iii)*
43. *iv)*
44. *i)*
45. *i)*
46. *i)*
47. *iv)*
48. *ii)*
49. *ii)*
50. *iii)*

www.ingramcontent.com/pod-product-compliance
Lightning Source LLC
Chambersburg PA
CBHW022024170526
45157CB00003B/1338